YOUR KNOWLEDGE HAS VALUE

Bibliographic information published by the German National Library:

The German National Library lists this publication in the National Bibliography; detailed bibliographic data are available on the Internet at http://dnb.dnb.de .

Imprint:

Copyright © 2019 GRIN Verlag
Print and binding: Books on Demand GmbH, Norderstedt Germany
ISBN: 9783346100399

This book at GRIN:

https://www.grin.com/document/511873

Kapya Tshinangi

Determination of the optimum conditions for the leaching of Co based on simulated data

Process optimization by using experimental design

GRIN Verlag

GRIN - Your knowledge has value

Since its foundation in 1998, GRIN has specialized in publishing academic texts by students, college teachers and other academics as e-book and printed book. The website www.grin.com is an ideal platform for presenting term papers, final papers, scientific essays, dissertations and specialist books.

Visit us on the internet:

http://www.grin.com/

http://www.facebook.com/grincom

http://www.twitter.com/grin_com

FACULTY OF ENGINEERING, BUILT, ENVIRONMENT & IT

Department of Industrial and Systems Engineering

BDE 780: Design and Analysis of Experiments Semester project:

Determination of the optimum conditions for the leaching of Co based on simulated data: Process optimization by using experimental design

By: TSHINANGI KAPYA FABRICE
Date of Submission: 03/11/2019

TABLE OF CONTENTS

INTRODUCTION

The treatment of Cu-Co ores by hydrometallurgical processes imposes a preliminary leaching step. However, the majority of Cu-Co ores encountered contain cobalt in the form of heterogenite with a fairly high proportion in the form of cobalt 3+. This form of cobalt has a weak kinetic during leaching and requires the presence of a reducing agent for its solubilization in a sulfuric medium.

In this work, we are interested in the formulation of a Pregnant Leaching Solution. There are many variables which affect the quality and production of the product, including time, percent solids, reducer concentration, tool materials, geometry of particles, etc. but for the model purpose we only consider two. The yield of this chemical process is being studied by taking in consideration two most important variables that are thought to be the pH and the reducing agent; tree levels of each factor are selected, and a factorial experiment is performed based on simulated `data, Process optimization by using experimental design. Consequently, companies are forced to operate by using the trial and error method. The optimization of controllable variables can make a considerable contribution towards solving the problem. At this point, we are more interested in Analyzing the data and draw conclusions under the following conditions:

1.*Response*: yield of cobalt

Factor 1: Reducing agent, with Levels 1(20 g/L), 2(30 g/L) and 3(40 g/L)

Factor 2: pH with Levels 1, 1.5 and 2

2. *Issue:*

- Which level has best effect on response (means effect)?
- Which level gives smallest variance.

3.*Hypothesis:*

- H_o1: $\mu11 = \mu12 = ... = \mu1a$
- $Ha1$: at least one $\mu1i$ is different
- H_o2: $\mu21 = \mu22 = ... = \mu2a$
- $Ha2$: at least one $\mu2i$ is different
- H_o3: Interaction effects the same
- $Ha3$: at least one μij is different

3. *Significance level*: $\alpha = 0.05$

4. *Confidence level*: $1 - \alpha = 0,95$

5. Power *of the test*: $1 - \beta = 0.5$

6. *P (Type II error)* $= \beta$

7. $\delta = 1.5$

8. a = 3, b = 3, for a power of the experiment equal to 0.5 and a confidence interval of 0.95, n = 3.

9. *total number of repetitions N = 27.*

We will Prepare appropriate residual plots and comment on the model's adequacy and see under what conditions we would operate this process, thus considerably reducing the number of tests needed. As a result, nowadays, this and similar methods have become the focus of interest for both academics and companies, with the goal of increasing production quality and operating with greater efficiency. In this study, the effects of leaching parameters on the recovery of Cobalt are investigated based on the 2 basic principles of experimental design; the model will be replicated 2 times in order to obtain an estimate of the experimental error and also to obtain a more precise estimate of the dependent variable which in this case is the recovery, with $\delta = 1.5$, a= 3, b = 3 we have obtained a value of n = 3 for a power of the experiment equal to 0.5; then the total number of repetitions is obtained, which in this case is equal to 27; another basic principle which is Randomization will be applied in the purpose of averaging out the effects of extraneous factors that are present. The analysis of variance (ANOVA) technique is carried out to investigating parameters which are statistically significant. Then the optimal plan is obtained.

As mentioned above, we are interested in the formulation of a Pregnant Leaching solution by using 2F3L factorial design to determine if this impacts the cobalt recovery based on simulated values; the main idea here is to determine if there is any indication that either factor influences cobalt recovery, Analyze the residuals from this experiment, verify if two factors interact, describe the differences in the effects of the different levels of cobalt recovery, which level of temperature we would specify in this work. The experiment is represented in Table 1. This orthogonal array is chosen due to its capability to check the interactions among factors.

1. BACKGROUND

1.1. DEFINITION OF LIXIVIATION

Leaching is a process of extracting a soluble component of a solid by means of a solvent. In other words, it is the dissolution of one or more valuable metals for their subsequent extraction. This chemical dissolution is done selectively with a minimum solution of impurities [Mutombo, 2007].

The choice of leaching's agent is a function of several factors which are essentially the following:

- The chemical characteristics of the materials to be leached;
- The cost of the reagent;
- The selectivity of the reagent with respect to the constituent that one wants to leach;
- The corrosive action of the reagent and its consequences on equipment and the environment;
- The possibility of regeneration of the reagent.

1.2. KINETIC MODEL OF LIXIVIATION

For a very long time we have not been concerned with measuring the speed of a reaction. It was limited to judging it very slow, slow, fast, explosive. Then following a classic approach in science, we tried to specify more. It is not unusual to find expressions like this: the reaction is completed after 10 hours or 2 days, for example.

Lastly, precise definitions have been introduced which account for precise and non-vague facts such as the "completed" nature of a reaction. The chemical kinetics were thus constituted.

The study of chemical kinetics was very fruitful because it led to the knowledge of the mechanism of the reaction; the metallurgist is no longer satisfied with the balance sheet, he makes every moment a precise state of the reaction and he often becomes capable of changing the course of the latter [Germain and Burnel, 1975].

1.2.1. STEPS IN REACTIONS.

Heterogeneous leaching reactions include the following steps:

- The diffusion of the reactant species towards the solid-liquid interface;
- Adsorption of the reactants at the solid-liquid interface;
- The chemical reaction at the interface;
- Desorption of reaction products at the interface;
- The diffusion of the products of the interface towards the solution.

It is therefore necessary to analyze the various stages in which a reaction takes place, it being understood that it is the slowest step that imposes the overall kinetics of the heterogeneous reaction; three cases are to be considered

The reaction is controlled by diffusion: when the reaction rate at the interface is greater than that of diffusion of the reactants towards the interface. In this case, the activation energy Ea is between 1 and 3 kcal / mole. This type of process is very sensitive to agitation, because it reduces the thickness of the boundary layer resulting in the increase of the reaction rate as shown in the relationship below:

$$Speed = \frac{D}{\delta} \times A(C - C_i) \tag{1}$$

With,

D: diffusion coefficient of the substance;

δ: thickness of the boundary layer;

A: interface area;

C: concentration of the reactant in the bath;

Ci: concentration of the reactant at the interface.

Temperature has a weak influence on diffusion-controlled processes as shown by the Stockes-Einstein equation:

$$D = \frac{RT}{N} \times \frac{1}{2\pi r \eta} \tag{2}$$

With,

D: diffusion coefficient of the substance;

N: number of Avogadro;

R: constant of perfect gases;

T: temperature in Kelvin;

r: grain radius;

η: viscosity of the fluid.

1.2.2. FACTORS LEACHING.

Leaching can be influenced by the following factors:

- The particle size: The properties of the dispersed states are related to the contact surface between the phases. As a result, a fractional phase is much more reactive than a massive phase, which can even cause explosions if the reaction is exothermic:
- The concentration of reagents involved: A diffusion-controlled liquid-solid process can become a chemically controlled process when increasing the concentration of reagents in the liquid phase.
- Density: For leaching, the density of the pulp(slurry) is an important parameter of dissolution. When the density is high, very concentrated solutions are obtained after leaching.

- Acidity: when leaching cobalt hydroxide, sulfuric acid also reacts with other hydroxides in the discharge. To limit its contamination reactions, we work with very dilute acid solutions.

Despite the low acid levels in the solution, the leaching reaction of cobalt hydroxide will proceed because cobalt and copper have a high affinity for sulfuric acid.

- Temperature: In many cases, thermal energy can cross the energy barrier. (case of chemically controlled processes);
- Agitation: in a solid-liquid reaction, stirring increases the rate of dissolution when the process is controlled by diffusion [Kapya. T, 2013].

2. OBJECTIVE(S)

Initially, the experimental design method is used to plan a minimum number of experiments. After identifying the working levels of the design factors and the main performance characteristics of the product under study, the method can be successfully applied to the leaching process. variations of the main leaching parameters and their interactions are investigated using orthogonal array technique. A statistical analysis of 'signal-to-noise' ratio follows by performing a variance analysis. After developing some special criteria, which depend on our performance objective, the optimal levels of the design factors are determined. Cobalt leach results will be discussed and commented.

3.FACTORS AND LEVELS

Although several factors may influence the study during this Project, some were not considered. In the case of this work, it has been suggested to us that 2 factors, each of which has 3 levels of variation. Table 1 presents the 2 factors selected for this project and their considered levels.

Table 1. Orthogonal matrix of 2 variables studied at
3 distinct levels

Symbol	factor	Levels		
		1	2	3
A	pH	1	1,5	2
B	Reducing agent	20	30	40

3.1 DESCRIPTION OF THE TESTS

Conducting leaching tests with 2 factors each having 3 levels is gigantic and often unrealistic when it comes to repeating the tests more than 4 times. To solve this problem, we thought it wise to use an experimental design with 2 replications.For design of experiments with two factors (pH, reducing agent) and three levels for each factor, the factorial design here is a standard L27 orthogonal array employed. Each row of the matrix represents one run. The factors and their levels are assigned in Table 2. Factors A, B are arranged in column 2, 3 respectively. The columns represented coded variables whose values correspond to the levels of the factor they are associated with, and the lines represent the experimental conditions of an experiment, that is, the levels taken during this experiment by each of the factors

For analysis of the results and optimization of conditions for setting the control factors, Minitab 2019 software is used. MINITAB Version 19 used for this project is the windows version software for Design and Analysis of Experiments.

Table 2. Orthogonal Matrix L27 (32)

run	factor A	B	run	factor A	B
1	1	20	14	1.5	30
2	1	20	15	1.5	30
3	1	20	16	1.5	40
4	1	30	17	1.5	40
5	1	30	18	1.5	40
6	1	30	19	2	20
7	1	40	20	2	20
8	1	40	21	2	20
9	1	40	22	2	30
10	1.5	20	23	2	30
11	1.5	20	24	2	30
12	1.5	20	25	2	40
13	1.5	30	26	2	40
			27	2	40

3.1.1. INTERACTION BETWEEN FACTORS

The notion of interaction between parameters is subjected to observations of the notion of parallelism between linear adjustment lines of various values of the signal / noise functional metric (S / N) for each controlled parameter [Nkulu, G., 2012].

4. PLANNED RISK AND SIMPLE SIZE ESTIMATE

The acid leaching process depends on many operating factors that have a direct influence on the quality of the product obtained. Whatever the field of study, the experimenter is always faced with the difficult problem of the optimal organization of his tests. It always seeks to find a compromise between, factors to consider, costs, deadlines and getting the right information on the results. ; as we explained in the introduction the model will be replicated 2 times in order to obtain an estimate of the experimental error and also to obtain a more precise estimate of the dependent variable which in this case is the recovery, with $\delta = 1.5$, a= 3, b = 3 we have obtained a value of n = 3 for a power of the experiment equal to 0.5; then the total number of repetitions is obtained, which in this case is equal to 27.

5. SELECTION, ALLOCATION OF MATERIAL

Having made these tests in the past based on the classical method, we are proposed to repeat the same materials because they remain the same as well even though it is a study based on simulated information

- Balloons, beakers, Buchner;
- Burette, test tube, Funnels;
- stemmed glass, magnetic stirrer (REMI MOTOR);
- electronic scale; Stopwatch;
- pH meter (METROM) with an electrode;
- potentiometer (ORION);
- vacuum pump for filtration, thermometer, thermostatic hotplate;
- stand, filter papers; a clamp; a wash bottle;

EXPERIMENTAL ASSEMBLY

To carry out these leaching tests, an assembly was carried out during our tests with the classical method and this assembly also applies to the experimental method. This assembly would consist mainly of the following elements: Beaker, pH-meter (METROM) equipped with an electrode, potentiometer (ORION), Thermometer.

let us remember here that the materials mentioned above were not used for our project because we relied on simulated data to do our studies. these informations are used only as guidance in that an actual experiment should be performed

Figure 1. Photograph of stirred reactor used for medium leaching
Reducing acid of trivalent cobalt

6. Execution and analysis of experimental results

The results of the twenty-seven leaching experiments are presented in Table 4. The tests were performed following the order of the matrix of experimental conditions implemented during the optimization.

6.1. ANALYSIS OF THE SIGNAL-TO-NOISE (S/N) RATIO

Cobalt recovery is measured experimentally for each combination of control factors using a factorial design. The optimization of the measured control factors was provided by signal-to-noise (S / N) ratios. In this study, it is the maximum that is desired to measure the effect of level of each parameter on the variability of the results obtained; Thus the largest values of cobalt recovery are very important for quality improvement of this project. For this reason, the ''larger-the-better'' equation was used for the calculation of the S/N ratio [Yuan et al, 2008]; and on the basis of the conclusions obtained we arrived at determining the experimental optimum guaranteeing a better result with less disturbance and this with a reduced number of tests. Table 5a shows the values of the S/N ratios for observations of the cobalt recovery.

Table 4. Factor Information

Factor	Levels	Values
A	3	1.0, 1.5, 2.0
B	3	20, 30, 40

Table 3. Factors Information and results of experiments

randomization		factors		recovery of cobalt		Blocking	Residuals	
StdOrder	RunOrder	A	B	Rep1	rep2	Blocks	RESI1	RESI2
14	1	1	20	92.07	91.605	1	0.07	1.247333
10	2	1	40	91.14	87.792	1	2.203333	-1.55867
15	3	1	30	94.86	94.581	1	1.61	1.490333
22	4	2	30	91.08	92.92	1	0.303333	2.564
21	5	2	20	90.09	89.18	1	-1.59	-0.80833
8	6	1.5	20	91.14	88.35	1	-1.21	-0.41733
21	7	2	40	94.86	92.535	1	1.59	-0.624
12	8	1	20	90.09	85.904	1	-1.91	-4.45367
16	9	1	30	92.07	95.511	1	-1.18	2.420333
1	10	1.5	30	90.16	92.92	1	-0.34333	1.186667
2	11	1.5	20	93.84	90.16	1	1.49	1.392667
18	12	1	40	86.49	88.35	1	-2.44667	-1.00067
9	13	1.5	40	91.14	91.605	1	1.24	2.542
7	14	1.5	40	87.42	87.792	1	-2.48	-1.271
20	15	2	40	94.86	94.395	1	1.59	1.236
3	16	1.5	30	89.28	93.93	1	-1.22333	2.196667
22	17	2	20	90.09	89.18	1	-1.59	-0.80833
20	18	2	30	89.18	86.45	1	-1.59667	-3.906
24	19	2	20	94.86	91.605	1	3.18	1.616667
5	20	1.5	20	92.07	87.792	1	-0.28	-0.97533
23	21	2	40	90.09	92.547	1	-3.18	-0.612
13	22	1	40	89.18	91.91	1	0.243333	2.559333
11	23	1	30	92.82	89.18	1	-0.43	-3.91067
4	24	1.5	30	92.07	88.35	1	1.566667	-3.38333
27	25	2	30	92.07	91.698	1	1.293333	1.342
6	26	1.5	40	91.14	87.792	1	1.24	-1.271
17	27	1	20	93.84	93.564	1	1.84	3.206333

Table 5a. Response Table for Signal to Noise Ratios

Level	A	B
1	39.21	39.20
2	39.11	39.15
3	39.20	39.18
Delta	0.10	0.05
Rank	1	2

Table 5b. Response Table for Means

Level	A	B
1	91.43	91.27
2	90.40	90.75
3	91.25	91.07
Delta	1.03	0.53
Rank	1	2

The results of the mean calculations of the S/N ratios obtained from the statistical analysis made using the Minitab software, relative to the cobalt leaching recovery of the experimental tests are given in Figure 2. From the results presented in the tables 3 and 5, it is possible to obtain a graphical representation of these results and to realize the influence of the factors on the response. A representation of these percentages in the form of a diagram of the main effects allowed a better exploitation of these results. Figure 2 presents the main effect diagrams of leaching tests based on simulated data.

The purely statistical notion that is used in several engineering domains compares the sizes or ranges of the parameters to highlight the influence or dominance on a given process (in the case of the present work it is leaching). Thus, in this statistical analysis, the parameter (factor) having the greatest influence on the recovery efficiency of the leaching is visualized in Figure 2.

From the above, we note that the order of importance of the parameters (factors), given by Δ in Tables 5a and 5b, starts with the factor A (Level 1, S/N = 39.21), factor B (Level 1, S /N = 39.20).

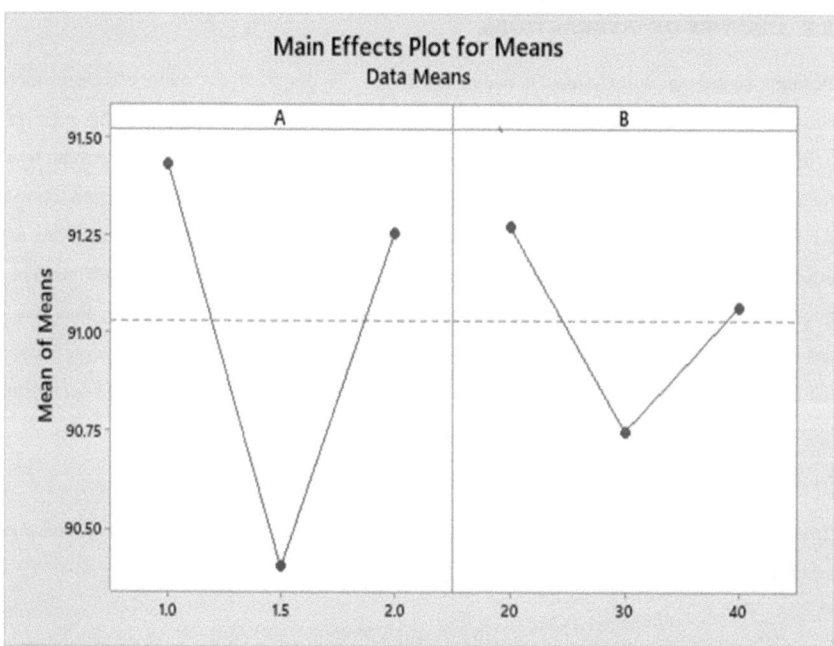

Fig 2. Graph of main effects of S/N of leaching

Examination of the results in the figure above shows that the reducing agent (factor B) has a less significant impact on cobalt leaching, thus improving its recovery. The pH (factor A) represents the most influential parameter on the 2 factors of our study. The order of importance of the most influential effects is as follows: pH, reducing agent. The optimal leaching conditions are as follows:

- reducing agent :20 g/L ;
- pH 1.

The optimal conditions provided by the main effects diagram of the S/N ratio are not necessarily those that give a better return. So we can assume that there is necessarily a synergy between the factors. Note that from the analyzed data, in the MINITAB output, the P values are greater than 0.05. Therefore, we reject the null hypothesis. It can therefore be concluded that there is a significant difference in levels.

6.2. ANALYSIS OF INTERACTIONS

Without significant interactions between the factors, a graph of the main effects cannot adequately describe the points where we can get the best results. The interaction in this exercise being significant, we must then study its diagram because a significant interaction between two factors can amplify or diminish the main effects and consequently influence the interpretation. An interaction is present if the answer for a level of a factor depends on the level or levels of other factors. In an interaction diagram, parallel lines indicate the absence of interactions. The greater the slope difference between the lines, the higher the degree of interaction. However, the interaction diagram does not indicate whether the interaction is statistically significant. (Minitab, 2016) In this part, we are interested in the interaction of the factors and the interaction hypotheses described below:

Null hypothesis: there is no interaction between the independent variables.

Alternative Hypothesis: At least one independent variable at one level is interacting with the other. level of significance of 5%.

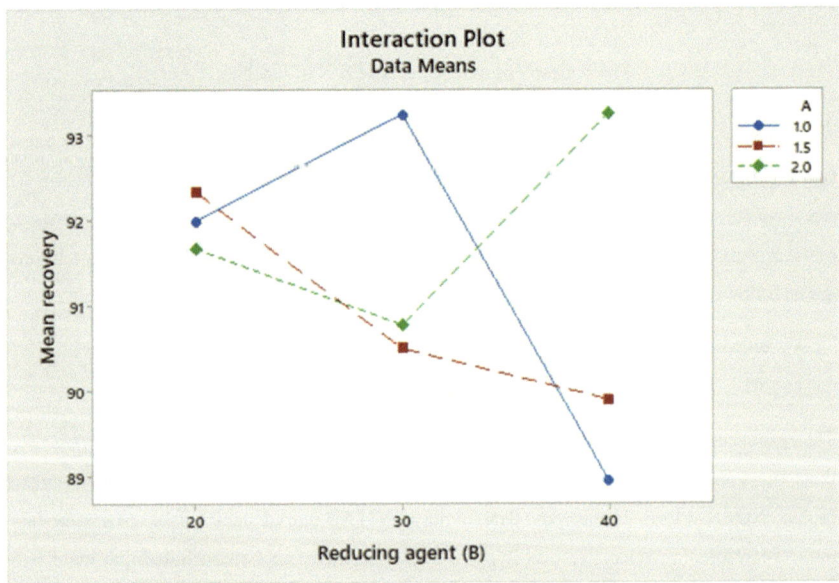

Fig 3. pH-reducing agent interaction diagram

The diagram of the pH (A) -reducing agent (B) interaction profiles as a function of the yield clearly shows that the interactions are not negligible and that they must be taken into account to make the interpretation.

The pH determines the dose of the reducing agent that has the highest level of yield and thus observing stronger areas of interaction than others in the field of study. More precisely, for a pH of 1, a yield of 92% is observed; with a concentration of the reducing agent of 40g/ml, when the reducing agent dose increases from 20 to 30 g/l, a 93.25% yield is observed for a pH of 1.5; finally, when the concentration of the reducing agent increases from 30 g/l. to 40 g/l a yield of 93.27% is observed for a pH of 2.

One thing that we have noticed is, from the MINITAB output provided in table 6, showing the P-value for the 2 way interactions, we have noticed that P-value is greater than 0.05. Therefore, by this condition, we reject the null hypothesis. Thus, it can be concluded that there is a significant interaction between independent variables.

6.3. ANOVA METHOD

ANOVA is a statistical method which is used to determine the individual interactions of all of the control factors in the test design. In this study, ANOVA was used to analyze the effects of pH, and the reducing agent on the recovery of cobalt. The ANOVA results are shown in Table 6. The analysis was carried out a 5% significance level and a 95% confidence interval. The significance of control factors in ANOVA is determined by comparing the F values of each control factor in the last column. The third column of the table shows the percentage value of each parameter contribution.

Table 6. Analysis of Variance

Source	DF	Seq SS	Contribution	Adj SS	Adj MS	F-Value	P-Value
Model	8	53.311	41.94%	53.311	6.664	1.62	0.186
Linear	4	12.261	9.64%	12.261	3.065	0.75	0.572
A	2	4.422	3.48%	4.422	2.211	0.54	0.592
B	2	7.838	6.17%	7.838	3.919	0.96	0.403
2-Way Interactions	4	41.051	32.29%	41.051	10.263	2.50	0.079
A*B	4	41.051	32.29%	41.051	10.263	2.50	0.079
Error	18	73.816	58.06%	73.816	4.101		
Total	26	127.128	100.00%				

By observing the above table, the recovery of cobalt, it is found that the percentages of contribution of factors A and B on the yield were found to be 3.48%, 6.17%. But, we have noticed that the interaction between the two factors has a yield yield of 32.29% on the yield.

6.4. REGRESSION ANALYSIS OF THE COBALT RECOVERY

Most regression analyzes are used for modeling and analyzing multiple variables when there is a relationship between the dependent variable and one or more independent variables (factors). In this project, independent variables are pH and reducing agent while the dependent variables is cobalt recovery. regression analysis was used to obtain a predictive equation for cobalt recovery; the predictive equation was only done for the linear regression model. The predictive equation is given below.

Regression Equation

Y = $91.407 - 0.012 \, A\_1.0 - 0.490 \, A\_1.5 + 0.501 \, A\_2.0 + 0.603 \, B\_20 + 0.103 \, B\_30$
$- 0.705 \, B\_40 + 0.002 \, A*B\_1.0 \ 20 + 1.752 \, A*B\_1.0 \ 30 - 1.754 \, A*B\_1.0 \ 40$
$+ 0.830 \, A*B\_1.5$
$20 - 0.517 \, A*B\_1.5 \ 30 - 0.313 \, A*B\_1.5 \ 40 - 0.831 \, A*B\_2.0 \ 20 - 1.235 \, A*B\_2.0$
30
$+ 2.066 \, A*B\_2.0 \ 40$

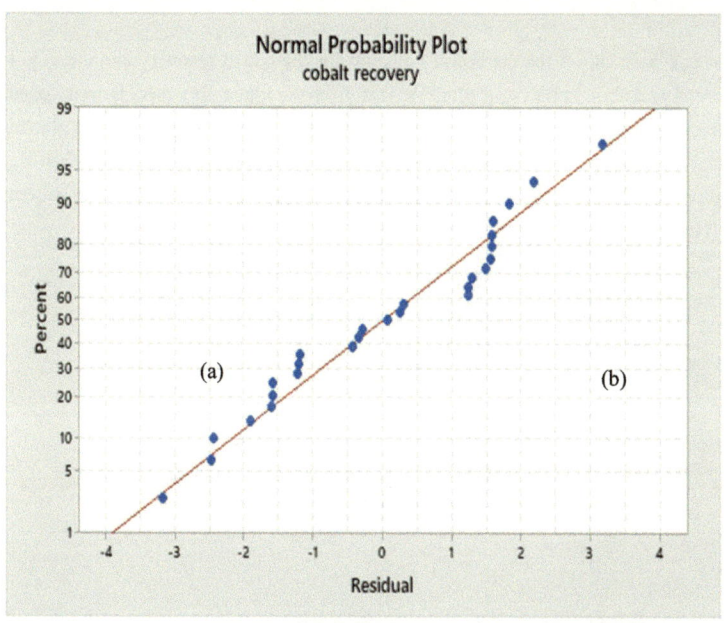

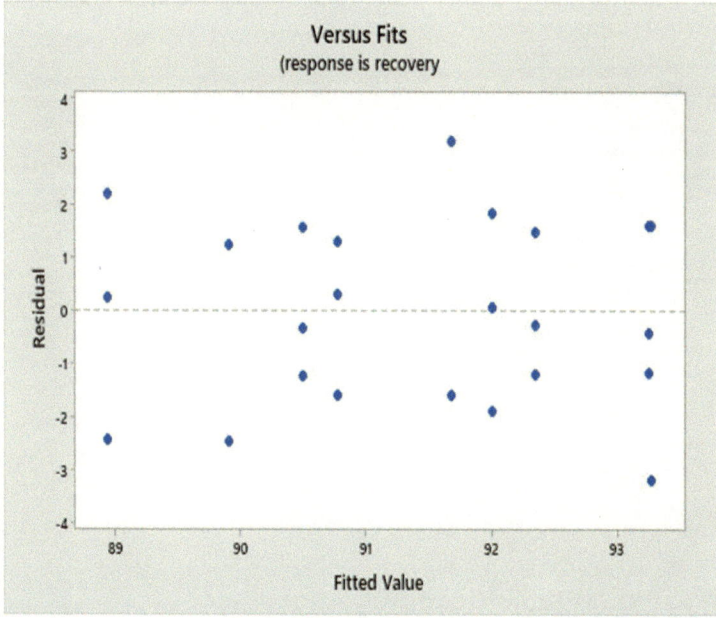

Fig 4. Normal probability Plot and residual versus fits

As we can see, Y here shows the predictive equation for the cobalt recovery and the Fig. 4 shows the normal probability plot vs fits plots. As seen from the figure 4(a), there is a very good relation between predicted values and test results. From the output, it is clear that the plotted residual values are closer to the line. Hence, the residuals are normally distributed. When we looked at figure 4(b), there seems some sign of problems but it is clear that there is no pattern in the graph. Hence, the line is a good fit for the data.

CONCLUSION AND PERSPECTIVES

The purpose of this work was to evaluate the influence of certain parameters on the recovery efficiency of cobalt during leaching based on simulated data. The factorial design methodology was used to better develop our tests and to process the results of the analyzes more efficiently. Two factors, varied at three levels, were chosen for the realization of this work, it is particularly: the pH, the reducing agent.

The study of the optimal conditions and the determination of the most significant factors for the improvement of the cobalt based on simulated data recovery was carried out using the technology of DoE. This technology has made it possible to study the effects of the variation of several factors as well as their interactions at the same time. The results of the experiments proved to be convincing and the evaluated techniques can be considered as promising. The results provided are as follows:

- The optimum levels of the control factors for maximizing cobalt recovery were determined. The optimal conditions for cobalt recovery were observed at A1B1;
- pH is the most influential factor on the 2 factors of the study and it was found that pH was the most significant parameter for the recovery of a percentage contribution of 91.43%.
- The optimal conditions according to the main effect diagrams, for the yield, obtained from the factorial design are: 20 g / L for the reducing agent and a pH of 1;
- a developed regression model demonstrated a good relationship with low correlation coefficients between the measured and predicted values for the recovery has been found since our data were not real but according to the observed test results, measured values were within the 95% confidence interval;

The diagram of the pH (A) -reducing agent (B) interaction profiles as a function of the yield clearly shows that the interactions were not negligible and that they must be taken into account to make the interpretation. Thus, it was concluded that there was a significant interaction between independent variables.

Given the results obtained on simulated data, we propose that if a research subject is to be carried out in practice, regardless of the type of domain, that all the parameters that can influence the process are taken into account to know which will be in blocking or in control in order to enrich the results of the data to be exploited, we also recommend to choose a better power of the expression for more reliability. Thus we can conclude by saying that this work is only a project to show how an experimenter should conduct his research, so we suggest that other methods of investigation must be used to provide additional useful results.

LESSONS LEARNT

DoE is one of the tools of quality. This tool is of interest only if we know how to use it, which supposes both knowing the method and being able to identify the cases where it brings a "more" compared to traditional techniques. The improvement of the quality of a product is conditioned by the implementation of a process which is less sensitive to the factors likely to affect this process. One of the things I learned in this module is that in quality engineering, the factors of a process are divided into two categories: controllable factors, that is, factors on which can easily act to control a specific process, and the uncontrollable factors, factors for which possible variations are not controllable in a process but may be the cause of instability in the performance characteristics of a system.

The study of the factorial design seen in BDE780 offers the possibility to better organize the tests that accompany a scientific research or industrial studies. With the plans of experiments, we could obtain the maximum of information with the minimum of experiments; the experimental design itself is a series of tests organized in advance in order to determine in a minimum of tests and with a maximum of precision the influence of multiple parameters on one or more answers. The choice the order of the essays will take into account these two aspects, some practical, others theoretical.

DoE has helped me to understand that the factorial design can help to design and industrialize a product, as well as solve complex optimization problems (adjustments) during product production. It fits perfectly in the quality approach.

REFERENCES

[1] Alvayai C.,«Hydrometallurgical treatment of Katanga cupro-cobaltiferous ores», Diploma of Advanced Studies, Faculty of Applied Sciences, University of Liège, 2OO6.

[2] Blazy P., JdidEl-Aid, «extractive metallurgy of non-ferrous metals», Mineral Industry Company, Saint-Etienne 1979, pp 40-45.

[3] Blazy P., JdidEl-Aid, «Copper hydrometallurgy », Engineering technique, Metallic materials 2003, M 2 242 pp 3-5.

[4] Bernard G., Roland G., «Chemical kinetics», TECHNOSUP, Ellipse, Edition Marketing

[5] Minitab software, 2016.A., 2004.

[6] Nkulu, G., 2012. Carrolite bioleaching - Applications to polymetallic sulphide ores from the Katanga Cupiferous Arc in DRC, Thesis, University of Liege.

[7] Yuan B., Xu L., Ren J. et Huang D., 2008. *Vibration Suppression Optimization of Electronic Apparatus Rackvusing Design of Experiment Method,* Asia International Symposium on Mechatronics, Hokkaido University, Japan, p.6.

[8] Mutombo, G. Non-ferrous extractive metallurgy course ", third graduate, UNILU polytechnique 2007.

[9] Twite, E " Course in metallurgical processes for special metals ", second grade, UNILU polytechnique 2009

[10] KAPYA TSHINANGI 2013 / Shituru Plants 2013